Microscopic Life
in the
Garden

BRIAN WARD

Contents and definitions

Bacteria

Bacteria are tiny organisms, that are far too small to be seen without a microscope. They may be rounded, thread-like, or rod-shaped, and some of them can creep around, using waving threads called **flagella**. Some bacteria need **oxygen** to live, while others are killed by oxygen in the air.

Viruses

Viruses are even tinier than bacteria. Unlike many bacteria, viruses must always infect a living cell if they are to grow and reproduce. They take over the whole of the workings of the cell and usually kill it as new viruses are released. This tomato plant has been infected by a virus.

Fungi

Fungi mostly feed on dead and **decaying** matter. Tiny fungi called "yeasts" grow naturally on the skin of many types of fruit, while others are responsible for breaking down dead plant and animal material in the process of decay. Some fungi cause plant diseases.

Algae

This is a group of tiny green plants that live on the soil surface or in damp or wet places. **Algae** look like tiny rounded shapes or green threads, and are common in puddles, gutters and ponds.

Creepy crawlies

We share our parks and gardens with millions of insects, spiders, worms and other creepy crawlies, some of which are very tiny. They all have an important role in keeping the soil healthy and in breaking down dead and decaying materials.

The healthy garden

Tiny microorganisms live in every park and garden. They perform many different tasks, from breaking down dead plants to preying on other bugs.

Microscopic life forms

Parks and gardens teem with **microbes**, many of which live in the soil where they cannot be easily seen. Some of these creatures play an important role in keeping the soil healthy; others cause plant diseases. Bugs in the garden are very dependent on temperature and water. In mild, damp weather they grow and reproduce very quickly, so plant diseases become more common.

Winter rest

In the cold weather of winter, the micro-organisms cannot grow and multiply as fast. The bugs stop much of their activity and most end up deep in the soil, protected from the cold weather.

Microscopic life in a garden is seasonal, like the growth of the plants it contains. Microorganisms are most active in the summer, but this activity stops during the cold winter months.

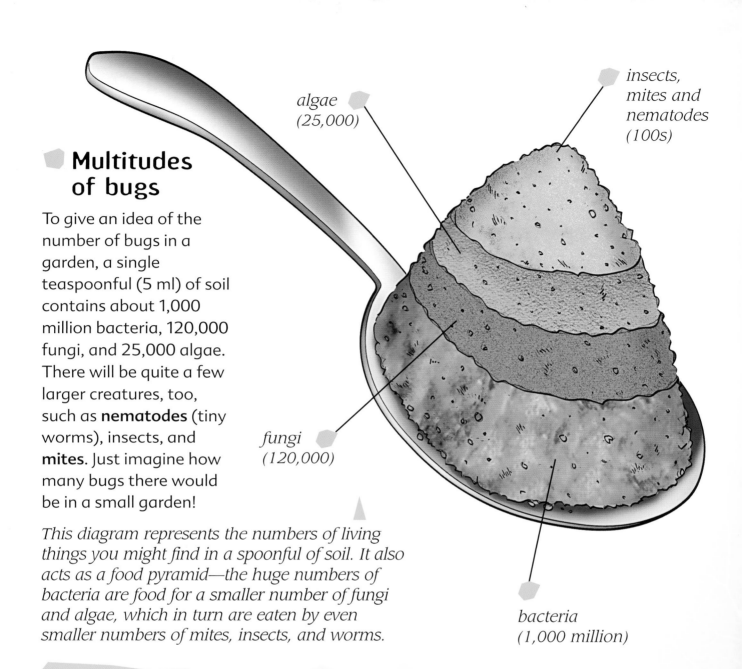

algae
(25,000)

insects,
mites and
nematodes
(100s)

Multitudes of bugs

To give an idea of the number of bugs in a garden, a single teaspoonful (5 ml) of soil contains about 1,000 million bacteria, 120,000 fungi, and 25,000 algae. There will be quite a few larger creatures, too, such as **nematodes** (tiny worms), insects, and **mites**. Just imagine how many bugs there would be in a small garden!

This diagram represents the numbers of living things you might find in a spoonful of soil. It also acts as a food pyramid—the huge numbers of bacteria are food for a smaller number of fungi and algae, which in turn are eaten by even smaller numbers of mites, insects, and worms.

fungi
(120,000)

bacteria
(1,000 million)

How many habitats?

There are lots of places where bugs can live in a garden, such as in the soil, on plants, or under dead leaves. How many places, or **habitats**, can you find in your garden or in your local park? Draw a simple map and mark out the dry, damp, shady, and wet places. Then try to work out which of the animals and plants described in this book might live in each place.

The food chain

The microscopic forms of life in parks and gardens feed on one another. This means that they are all closely linked with one another in a network called the **food chain**.

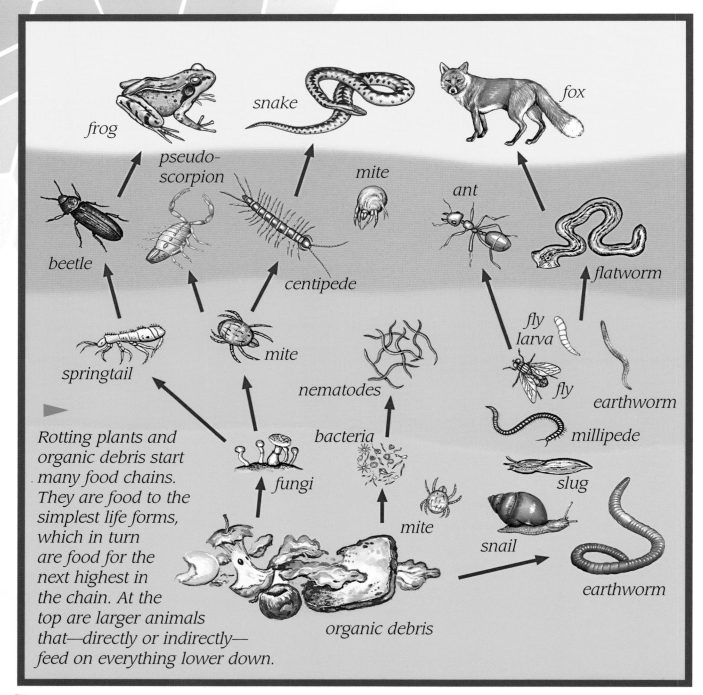

► Rotting plants and organic debris start many food chains. They are food to the simplest life forms, which in turn are food for the next highest in the chain. At the top are larger animals that—directly or indirectly—feed on everything lower down.

frog

snake

fox

pseudo-scorpion

mite

ant

beetle

centipede

flatworm

mite

fly larva

springtail

nematodes

fly

earthworm

bacteria

millipede

fungi

slug

mite

snail

earthworm

organic debris

What is at the bottom of the food chain?

All garden life depends on other forms of life for food. Green plants have an important place in the food chain. They use energy from the sun to make sugars (food) in a process called **photosynthesis**. The plants, in turn, become food for all sorts of animals, fungi, and even viruses. When plants die and decay, very simple organisms such as bacteria, soil algae, and tiny fungi feed on the dead or decaying material. They break it down into simple substances called "nutrients" that can be absorbed easily and used to feed plants. Many small animals feed on the tiniest bugs.

▲ *A wide range of animals find their food in organic waste, aiding the process of decay.*

Higher up the food chain

In the garden a large part of the food chain takes place deep in the soil. But examples can also be seen above the soil. Leaves fallen from trees rot and are often eaten by worms before microorganisms break down the tiny matter that is left. Plants are eaten by animals such as worms, slugs, and snails, which in turn are eaten by birds and rodents.

MICRO FACTS — Top of the chain

Because we can eat all sorts of animals and plants, while virtually nothing preys on us, humans are at the top of the food chain. When we die, our bodies can enter the food chain again through the process of decay.

Rot and decay

Fungi are one of the most important links in the food chain. These organisms are responsible for breaking down most of the plant material and dead bugs in the garden, releasing their nutrients to be used by other living organisms.

__Actinomycetes__ are a very common organism that breaks down plant waste. Their tiny threads grow through the waste, releasing __enzymes__ that break it down.

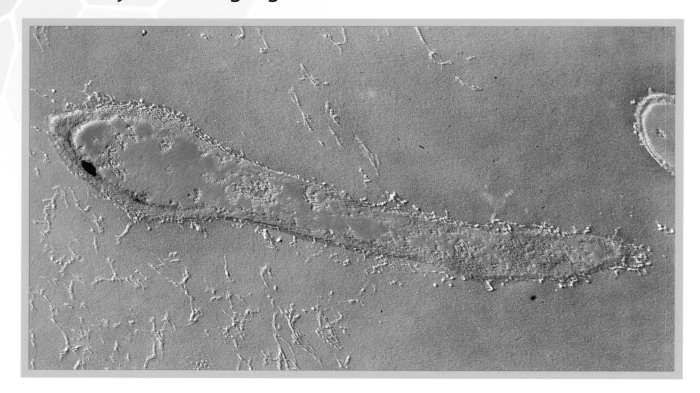

Starting to rot

Without fungi and bacteria, our gardens and the whole countryside would be covered by a deep layer of dead plant material. But rot caused by these organisms makes it possible for all of the valuable food substances locked up in dead plant material to be returned to the soil to fertilize it.

Thread-like chains of bacteria called "Actinomycetes" grow through litter very quickly. Their main job is to attack woody plant remains. These bacteria can be seen if a compost heap is disturbed. They form a powdery white layer just below the surface of the litter.

Fairy rings

Fungi reproduce by tiny **spores**. In some types, the spores are carried on structures called "fruiting bodies," which we know as mushrooms. These often appear in lawns and sometimes grow in a circle called a "fairy ring." The rings grow slowly from a central point, and as the fungi die off, they release nutrients. These make the grass grow more strongly, so that a green ring appears even when no mushrooms can be seen.

Fairy rings appear as fungi grow in a circle in the grass.

These microscopic fungi produce their spores from rounded organs that are raised above the damp surfaces on which they live.

MICRO FACTS   ## Be careful!

Never eat wild mushrooms unless an adult has checked them and has said that they are safe to eat. Some types could make you very ill.

Plant food

Some fungi called "yeasts" are simply tiny single cells, but most are multi-celled and grow in thread-like strands called **hyphae**. They grow through the soil and cover dead plant material. The hyphae produce substances that break down the dead matter and make the food easy to absorb. As the plant rots, different types of fungi take over. The substances produced from rotting plant remains are an important source of nutrients for other green plants.

Compost and recycling

People produce huge amounts of waste, and much of this is dumped into pits called" landfill sites," which is an expensive way to clear up litter. Fortunately, a lot of garden and household waste can be treated more sensibly—by composting.

Starting the compost heap

Composting is the way to use the natural process of rot and decay to break down garden waste into harmless substances. These can be added to the soil to help garden plants to grow better. Composting puts the food chain to work for people. It is very easy to start the process. Garden waste and vegetable waste from the home (including paper, which is made from wood) are piled up in a convenient corner of the garden, or put into a recycling or composting bin, and kept moist.

*Vegetable garden waste can be placed into a **compost** heap, where the process of decay takes place.*

What is in the compost?

There are tiny relatives of scorpions living in your compost heap! These are called **pseudoscorpions**. They are only slightly larger than a pinhead and are very shy. They have crab-like pincers like real scorpions, but they do not have a stinger. They are only dangerous to the tiny insects that they eat.

Breaking down the litter

Within a day or so, billions of bacteria and fungi get to work, breaking down the plant waste. At the same time, this process produces a lot of heat, which encourages the growth of different types of bacteria. After a few days they have done their work; then they are replaced by a host of other tiny creatures that feed on the litter.

How does it feel?

Put a few large handfuls of fresh grass cuttings into a plastic bag with no holes in it. Close it tightly with an elastic band. Feel the bag every day to see how hot it becomes. When you are finished, throw the bag away without opening it.

Inside the compost heap

Common residents eating their way through the litter include **springtails** (tiny insects that jump when they are disturbed) and millions of spider-like mites. Tiny worms called "nematodes," earthworms, grubs, and slugs also play a part in breaking down rotting waste materials. These animals are all preyed upon by **predators** such as centipedes and beetles.

Mites are tiny relatives of spiders. They grow in huge numbers in a compost heap.

A rich compost

Eventually, the waste is broken down to a harmless, but rich, compost. When spread on the garden, it helps the plants to grow well and improves the soil condition. Compost should not be made from plants that have been recently sprayed with garden chemicals. These chemicals could build up to levels that could harm other plants when the compost is spread on the garden.

Fungal enemies

Fungi can be very useful in breaking down waste, but some can also be a menace to the health of plants, and may even kill trees.

Tree killers

In the last 40 years, millions of elm trees in North America have been wiped out by **Dutch elm disease**. This disease is caused by infestations of beetles burrowing into the bark of elm trees, allowing fungi to get in and choke the water channels. The trees are starved of food and die. Dutch elm disease is thought to have come from Asia. It has already killed almost all of the elm trees in Western Europe.

These elm trees have been left dead and leafless by fungal disease.

A fungal disease called "honey fungus" is common in gardens, usually attacking trees that are very old and already diseased. The first sign of infection is clumps of yellowish mushrooms growing near the base of the tree. By the time they appear, it is too late to save the tree because tiny threads of hyphae will have already penetrated the roots and trunk.

Yellow mushrooms growing on or near a tree are a sure sign that it is being attacked by lethal honey fungus.

Rose diseases

Lots of different fungi attack other garden plants. Roses are very likely to get **mildew**. This produces powdery white patches on the leaves, causing them to shrivel up. Other fungi cause black patches on the leaves of roses. Similar diseases caused by fungi infect other types of plants. Many of these are of a type called **rusts**, which produce orange or brown stains on the leaves.

Roses are often attacked by mildew, a fungus that produces powdery patches on the surface of the plant.

Healthy or diseased?

Seedlings are often attacked by a fungal disease called "damping off." Soil fungi that cause damping off live in moist conditions. They look like tiny, fluffy grey threads that may be seen moving across the soil. The fungi penetrate the stems of the seedlings and block the water channels, so the young plant becomes limp and collapses. Fill two plant pots with compost. Sprinkle mustard and lettuce seeds on top and water them. Now stand one pot on the windowsill and keep it watered just enough to keep the surface moist. Seal the other pot tightly in a clear plastic bag with no holes. The seeds in both pots will sprout quickly. Look every day to see how the plants are growing. Do they look the same, or does one pot look healthier than the other? Why do you think this is?

Plant viruses

Viruses are very unusual organisms because they cannot live and reproduce on their own. Instead, they have to take over another living cell.

▲ *Virus infection is causing the leaves of this tomato plant to curl. The leaves will become yellow and die.*

Getting into the plant

All viruses are **parasites**, surviving on other living things and harming them. They need to get inside a living cell so that they can begin to produce more viruses. They usually infect plants when plant-sucking insects such as aphids puncture the plant's surface, allowing the virus to enter a cell. Other organisms that spread viruses include nematodes, fungi, and mites. Once in the plant, viruses spread rapidly. Unlike animals, plants do not have an effective immune system, so infected plants may die.

Signs of attack

When plants have been attacked by a virus, yellowish patches appear on their leaves. Viruses often attack tomato plants, either in a greenhouse or outdoors. They produce yellowish patches on the plant's leaves, which then shrivel up. Other viruses make plant cells reproduce too quickly, producing lumpy and misshapen leaves and shoots.

Virus control

It is very difficult to fight viruses because they are protected inside the plant's cells. The only way to keep virus diseases under control is to remove the infected parts of the plant and destroy them to prevent the disease from spreading. Spraying plants with **pesticide** can not only control aphids and other biting insects that help spread the virus, but also reduce the extent of the infection.

Colourful tulips

Tulips with brilliantly colored stripes and patterns on their petals were once unusual and very expensive. It was discovered that the bright patches of color on these tulips were caused by a virus infection that affected the way the plants developed. The virus often killed the plant. Today, modern tulips are bred with the same bright colors, but they are free from the virus.

Brightly striped tulips like these were once produced by a virus infection.

Friendly bugs

Many insects and other bugs threaten the health of plants in our gardens and food crops in the countryside. Highly poisonous chemicals can be used to control them, but we can also use nature's own way to combat these pests.

What is biological control?

Biological control means using natural predators to kill the bugs that damage plants instead of using poisonous chemicals, which would kill the "good" bugs, too. For instance, ladybugs can be used for biological control. These beetles feed on aphids, and so do their grub-like **larvae**. Spraying the aphids would kill the ladybugs as well, so a new crop of aphids would soon develop.

TRY IT YOURSELF

Garden friends

In spring or summer, look for tiny aphids on the shoots of plants in your garden. Then look carefully for the ladybugs that feed on them. You may even see the greenish grubs that are the ladybug larvae, which feed even more greedily on the aphids.

Ladybugs are familiar small beetles. They and their larvae eat huge numbers of aphids and other pests.

Bugs by post

Most predators used to control pests are insects such as wasps, flowerflies, and lacewings. Usually, they attack only a particular type of pest. Gardeners can buy these predators by mail order. They are sent out in small, sealed packages that contain eggs or larvae, which are then released on the plants. Biological control is usually used in greenhouses, where the useful predators can be prevented from escaping.

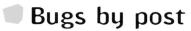

"Friendly" predators can be bought by mail order and released to protect crops. This avoids the need for powerful chemical sprays.

Encouraging disease

Predators kill pests by feeding on them, but a different type of biological control can be carried out using organisms that cause disease in the bugs. It is not just insects that can be used to control pests in this way. Special cultures of bacteria can be sprayed to kill bugs. *Bacillus thuringiensis* is a poisonous bacteria that infects and kills insects, caterpillars, and slugs. Fungal spores can be sprayed onto plants. The *Verticillium lecanii* fungus attacks whitefly adults and larvae, but is harmless to plants and "good" predators.

Caterpillars infected with Bacillus thuringiensis *do not become butterflies. Instead, they become swollen with bacteria and eventually split open, releasing billions more bacteria to float in the air and infect other caterpillars.*

17

Plant pollen and fern spores

Outdoors, every breath you take contains thousands of tiny plant cells. These are either pollen grains shed from flowering plants, or spores produced by ferns.

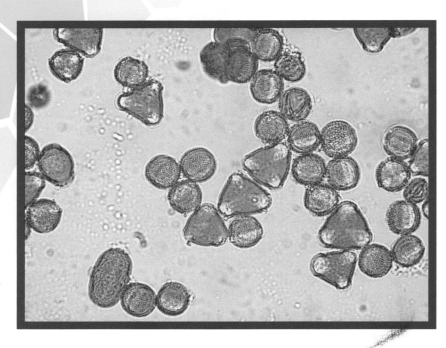

*On a warm summer's day, the air is filled with microscopic **pollen** grains.* ▶

⬡ Floating pollen

People who brush a hand over a flower may find a yellowish powder coating your hand. This is pollen. The grains are almost too small to see—less than the width of a human hair—but they are often brightly colored. They are so light that they are carried by the breeze to fertilize other flowers, so that these flowers can reproduce.

⬡ A risk to health

A single plant can sometimes produce a million pollen grains each day. When you breathe in lots of pollen grains, substances on their surface are sometimes mistaken by the body for dangerous invaders, so the body reacts against them. This can set off an "allergic reaction."

⬡ Pollen allergy

Lots of people have an allergy to plant pollen, which causes hay fever. People suffer from wheezing and runny eyes when certain plants are in flower. Grass pollen is the most common cause of hay fever, which usually occurs in early summer. Trees such as birches release their pollen in the spring and this, too, can cause allergic reactions.

◀ *Millions of pollen grains are produced on flower organs called "anthers."*

Fern spores

Fern spores also float in the air. Ferns do not produce flowers. Instead, they produce tiny spores on the underside of their leaf-shaped fronds and use these to reproduce. Fern spores, too, can sometimes bother people with allergies.

Tiny fern spores are produced in organs on the underside of the fern's leafy fronds.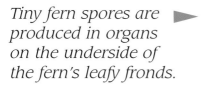

TRY IT YOURSELF

Look for fern spores

Find a fern—they usually grow in damp, shady corners of a garden or park—and look under one of the larger fronds. Can you see the raised bumps on it? This is where the spores are produced. Cut off the frond and lay it face up on a sheet of white paper. After a week, lift the frond carefully, and you will find an attractive pattern of dark spores on the paper. A large fern frond can produce about a million spores!

Lichens

Gray or yellowish patches can sometimes be seen growing on roofs or rocks. This is lichen, and it is a very strange form of life. Lichens are not plants. They are actually a mixture of a fungus and either algae or blue-green colored bacteria living inside it.

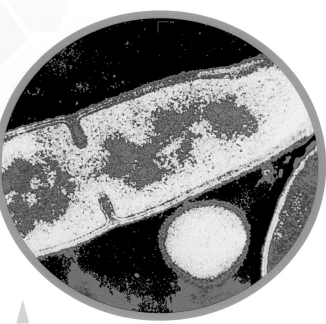

*Tiny cells of algae live inside a fungus to produce **lichen**. The fungus provides shelter for the algae, which in turn produce food for the fungus.*

Grayish-green lichen often grows on trees in damp places. Other types grow on rocks or even on rooftops.

Strange life

This odd mixture of fungus and algae or bacteria means that lichens are able to survive in places where little else can live. The fungus provides protection for the algae or bacteria, while they make food for the fungus, using sunlight. The result is a partnership between two very different forms of life, which allows lichens to live on very little nourishment. This type of partnership is called **symbiosis**. Lichens do not need much water because they can absorb moisture from the air. This is why they can live in exposed and unwelcoming places.

Reindeer moss

Some lichens grow into the shape of a tiny, branched shrub. Lichens of this type form the main food for reindeer that live on the Arctic plains known as tundra. Lichens grow on fence posts and trees in sheltered damp spots. The branched types grow mostly among damp moss in shady woods.

Lichen is the main diet of reindeer. It is known as "reindeer moss."

Pure air

Lichens grow very slowly and live for many years, forming a thin crust over roofs, rocks, and tree branches. Because lichens depend on moisture in the air, they are sensitive to air **pollution**. Some are killed more quickly than others, so scientists are often able to detect air pollution by studying the different types of lichens. In towns, only the toughest forms survive; they can be seen growing as gold or gray patches on rooftops.

Useful lichens

Lichen was once used to make dyes for clothes, but now their only common use is to make a dye called **litmus**. This dye is produced from lichen grown in the Netherlands for use in chemistry laboratories. It changes color when mixed with solutions of acid or alkaline substances.

TRY IT YOURSELF

Acid or alkaline?

Ask your teacher for some strips of litmus paper. The paper will turn red when dipped into acid liquids and blue when dipped in alkaline liquids. Try dipping strips of paper into lemon juice, vinegar, liquid detergent, dissolved detergent powder, and lots of other liquids to see if they are acid or alkaline.

Helpful bugs

Farmers and gardeners add fertilizer containing nitrogen to the soil to make the plants grow faster. Most of the air we breathe is made up from this colorless and harmless gas, so why does it need to be added artificially?

*Gardeners often add fertilizers containing **nitrogen** to the soil in order to encourage plant growth.* ▶

▲ *Bacteria growing in nodules on the roots of some plants "fix" nitrogen from the air as a form of natural fertilizer.*

Nitrogen from the air

Nitrogen is a chemical that encourages plant growth, but most plants cannot absorb it from the air. They take nitrogen from the soil, where it is produced when fungi and bacteria break down dead plant material. Nitrogen is often added to the soil by gardeners in the form of chemical fertilizers. But some plants have another way of obtaining nitrogen, using "friendly" microbes. One of these is a bacterium that lives in small swellings on the roots of plants such as peas and clover. This bacterium can "fix" nitrogen from the air. It uses some itself, and the rest is used by the plant.

Fungal helpers

Tiny threads of a fungus called **mycorrhiza** grow into the roots of most trees and many other plants. These threads spread out around the plant, absorbing food substances that the plant needs but cannot take in by itself. The fungus breaks down food substances in the soil and feeds most of them back to the plant. In return, the plant supplies the fungus with the sugar it needs for growth. These fungi allow plants to grow in many places where few nutrients are available. Mycorrhiza is very important to the growth of healthy trees and plants, so if it is missing from the habitat farmers may add it to the soil.

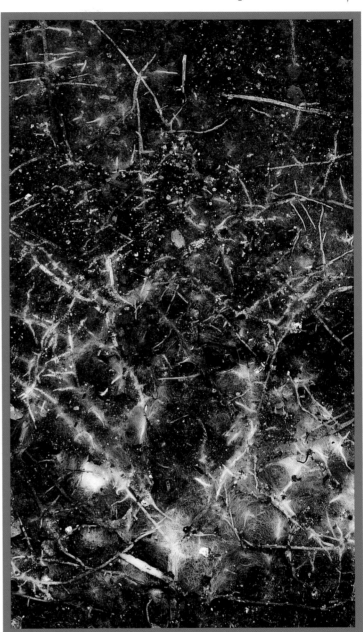

Mycorrhizae are fungal threads that spread out in the soil, passing food substances back to the plant roots. ▼

Natural killers

Many other useful bacteria live in the soil, including some that can be used as biological pesticides to kill insects. There are others that contain substances from which **antibiotic** drugs can be made.

23

Life in pools and puddles

Microscopic life teems in every drop of water in the garden. This may be a puddle, a blocked gutter, or a pond. Almost all forms of garden life can be found in or near water.

What is in the water?

Bacteria live anywhere, including in water. Bacterial decay is responsible for the nasty smell of stagnant water. Algae are tiny plants that also grow in huge numbers in water. They convert sunlight to sugars, using the green **chlorophyll** inside them. Some algae are single cells that make the water look green. Other types of algae form long, green threads, which are often seen in ponds. This is known as "blanket weed." Blue-green bacteria grow in similar long threads, usually in smelly, polluted water.

Billions of microscopic forms of life swarm in the layer of scum on a pond or puddle.

Aquatic animals

Tiny single-celled animals feed upon the bacteria and algae swimming or gliding in water, and they in turn are eaten by larger creatures called "rotifers." Rotifers are animals that have vibrating hairs round their mouths, which create water currents to carry their prey straight into their stomachs.

Rotifers are sometimes called "wheel animals," because of the vibrating hairs they use to catch prey, which look as if they are going around and around.

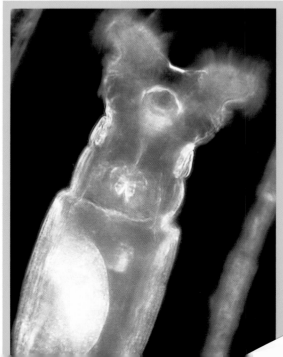

Aquatic food chain

In an underwater food chain (see pages 6 and 7), all these tiny forms of life provide food for slightly larger animals. Many are eaten by hydra, which are tiny animals shaped like a sea anemone. Hydra catch their prey with stinging tentacles and then stuff it into their mouths. Tiny flatworms called **planaria** glide about, eating a film of bacteria and algae. Larger worms burrow through the mud, feeding on dead material.

Still larger creatures, such as midges, lay eggs in the water, and their larvae hatch out and swim about. These in turn are eaten by large water insects, fish, and frogs. Finally, the larger water-dwellers are themselves eaten by predators such as water birds, snakes, muskrats, and mink, which are at the top of the aquatic food chain

Tiny planaria glide across ▶ *underwater leaves, grazing on bacteria and other small organisms.*

TRY IT YOURSELF

Watch them swim

Scoop two large spoonfuls of garden soil into a jam jar. Fill the jar with water and screw on the lid or seal it in a plastic bag to prevent airborne bugs from getting in. Leave it on a windowsill where it will get plenty of light. Look at it carefully every day. Soon you will see green patches on the glass as algae grow. The water may turn green, too. After a while you will probably be able to see tiny creatures swimming in the water. All of these organisms will have developed from creatures, eggs, and spores in the soil.

Nematodes, insects and other bugs

Lots of creepy crawlies live in the soil and cause damage to plants, but most of them play a part in keeping the garden healthy and cause us no harm.

The soil is full of tiny nematode worms. Their length is no greater than the thickness of a penny.

Tiny worms

Nematodes are tiny worms that exist in huge numbers in the soil. A few of these nematodes damage plants, but they have many other sources of food. Some eat bacteria, or suck the juices out of fungi or algae in the soil. Others attack insect pests, keeping down their numbers naturally. By recycling food materials and spreading them through the soil as they move about, nematodes play an important part in the food chain They are also eaten by larger predators. Several hundred nematodes could be found in one teaspoonful (5 ml) of garden soil, and as many as 90,000 in a rotting apple. There could be millions in a handful of material from a compost heap.

Jointed legs

Soil contains a huge number of **arthropods**. These are animals with jointed legs, such as insects, mites, centipedes, and millipedes. There could be 90,000 or more arthropods in a square yard (.8 sq m) of garden soil, and most of these would be springtails. These tiny insects have a jumping organ folded underneath them; when disturbed they shoot it out and catapult into the air. Springtails feed on decaying material and play an important role in the breakdown of dead matter. By chewing up woody plant material they make it easier for

Springtails are very common in compost heaps, but they are also found on the surface of pools of water. They jump high into the air when disturbed.

bacteria and fungi to break down the waste, and so speed up the process of decay. Mites break down plant waste in the same way—there are hundreds of different types of mite in the soil.

TRY IT YOURSELF

What lives in the soil?

Take an empty plastic bottle and throw away the top. Make a funnel by cutting the bottle in half. Turn the neck part upside down and stand it in a cup containing a little water. Place a nylon scouring pad in the funnel and press down. Put a couple spoonfuls of garden soil on top of the pad (which stops the soil from falling through the funnel). Place the funnel and cup about four inches (10 cm) below the light from a reading lamp, and leave for a few hours so the heat from the bulb warms the soil. The organisms in the soil will move away from the heat, pass through the pad, and fall into the cup, where you can look at them with a magnifying glass.

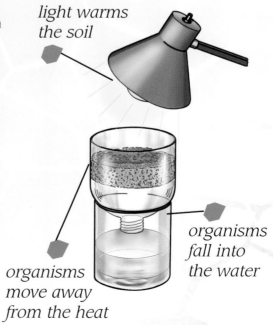

light warms the soil

organisms fall into the water

organisms move away from the heat

Health hazards

Lots of fresh air outdoors is good for people, but there are some unexpected dangers in a garden.

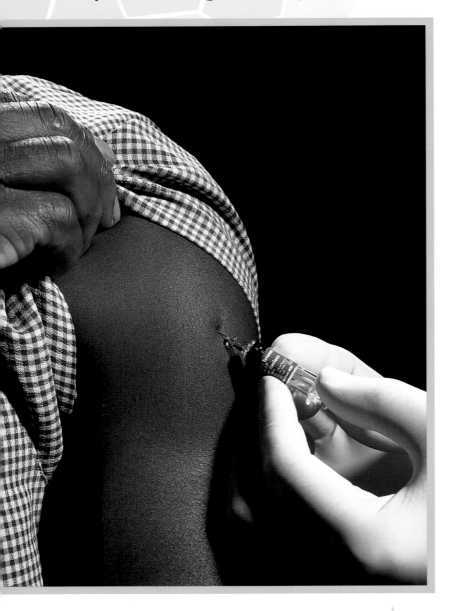

*Injuries in the garden can allow dangerous **tetanus** germs to enter the body. After cleaning the wound, a protective vaccination is often given.*

How bacteria can infect people

Cuts and scratches always carry the risk of infection. Soil and plants are covered with bacteria, and some are able to enter the body through cuts and splinters, causing a nasty infection that must be treated with an antibiotic. A wound has become infected if it is red, warm, and swollen. If this happens, prompt medical attention is needed. Cuts and scratches should be washed and kept covered until they begin to heal.

Septic wounds

Deep and dirty wounds can be particularly dangerous, because sometimes soil contains bacteria that can cause the very serious disease called "tetanus." For a large wound, the doctor will often give an injection to protect against tetanus. This protection will last for a long time. In many countries, people are routinely vaccinated against tetanus.

Painful plants

Stinging nettles have a microscopic means of defence. They are covered with tiny tubular hairs containing poison. When you brush against them, poison is injected through the skin, producing aching and sometimes blistery swellings that are painful for a while. Similar hairs are found on lots of garden plants. They may not sting, but many of them, such as the hairs on primroses, may cause an itchy rash.

Warning: Never eat a plant or berry from the garden unless an adult says that it is safe.

The stems and leaves of nettles are covered with sharp, glassy needles that inject poison into the skin when touched.

TRY IT YOURSELF

Avoiding mosquitoes

Stagnant water in an old bucket or pail may be a breeding place for mosquitoes that will bite you if you are in the garden in the evening (when they are most active). They carry diseases such as **malaria** and West Nile virus.

A mosquito lays its eggs in clusters on the surface of some stagnant water.

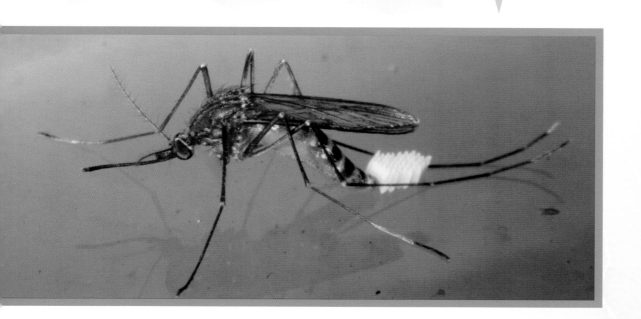

Glossary

Actinomycete: Thread-like bacteria that often live in compost heaps.

Algae: A type of plant with a very simple cell structure.

Antibiotic: Drug that attacks bacteria and sometimes other microbes. Many antibiotics are produced by molds living in the soil.

Arthropod: Animal with jointed legs, such as an insect, a spider, or a centipede.

Bacteria: Tiny single-celled microbes that live nearly everywhere, including gardens, homes, and humans. They may enter plants and cause disease. They have a tough outer coat.

Chlorophyll: Green substance present in most plants. It uses sunlight to convert other substances in the plant to sugar, which it uses as food.

Compost: Material remaining after plant matter has decayed. Compost is used to improve the structure of soil.

Decaying: Being broken down by microbes.

Dutch elm disease: Fungal disease that kills elm trees. It is caused when burrowing beetles allow the fungus spores through the tree bark.

Enzyme: Substances produced by living cells that help a chemical reaction to take place.

Flagella: Thin hair-like strands found on some bacteria and other microbes, which allow the microbe to move about.

Food chain: System in which one form of life eats another. It is itself eaten by another creature further up the food chain.

Fungi: Organisms that break down dead material and can sometimes cause disease. Most fungi are microscopic; others are large, such as mushrooms.

Habitat: The place where organisms live.

Hyphae: Tiny threads of fungus that penetrate the soil or food materials.

Larvae: Young forms of an animal that do not look like the adult. For example, a caterpillar is the larva of a butterfly.

Lichen: Primitive organisms that are combinations of fungi and algae or bacteria. They grow very slowly.

Litmus: Purple dye obtained from lichen, that changes color in the presence of acids or alkalis.

Malaria: Tropical disease caused by the bite of a mosquito. It is common in parts of Africa and Asia.

Microbe (microorganism): Microscopic organism.

Mildew: Disease of plants, particularly the leaves, caused by fungi.

Mite: Tiny spider-like animal that feeds on plants and decaying matter in the soil.

Mycorrhiza: Fungal thread that penetrates the roots of some plants, allowing the plant to absorb nutrients.

Nematode: Tiny thread-like worm that is very common in the soil and in compost heaps.

Nitrogen: Colorless gas present in the air, which helps plants to grow.

Oxygen: Colorless gas in the air we breathe.

Parasite: An animal or other living organism that feeds on other forms of life.

Pesticide: Substance that kills garden pests.

Photosynthesis: The way plants use energy from the sun to turn carbon dioxide in the air and water into food.

Planaria: Tiny flatworms, common in water and in damp soil.

Pollen: Tiny grain produced by a flower that fertilizes an egg cell of the same variety of flower so that a seed can form.

Pollution: Contamination of air or water. Factories and car exhausts cause air pollution; sewage can cause water pollution.

Predator: Form of life that feeds on other living animals.

Pseudoscorpion: Very small arthropod with pincers like a scorpion but without a stinger.

Rust: Fungal disease that produces discolored patches on plant leaves.

Spore: Single-celled bodies that are the means of reproduction for fungi, some bacteria, algae, and ferns.

Springtail: Tiny insects that flip into the air when disturbed. They are common in the soil.

Symbiosis: Situation in which two different forms of life benefit from living together. Fungi and algae live together in symbiosis in a lichen.

Tetanus: Dangerous disease caused by bacteria living in the soil, which can enter the body through an open wound. Tetanus can be prevented by vaccination.

Virus: Very simple organisms that can grow and reproduce only inside living cells. All viruses are parasites.

Further information

The following web sites contain lots of useful information about microbes and their effects on the garden:

Biological Control of Pests:
**http://smartgardening.com/
Biological_Pest_Control.htm**

Compost Biology:
**http://smartgardening.com/
Biology_of_Backyard_
Composting.htm**

The Friendly World of Lichens:
**http://www.earthlife.net/
lichens**

Fun Facts about Fungi:
**http://www.herb.lsa.umich.
edu/kidpage**

Ladybugs:
**http://www.viterbo.edu/
academic/ug/education/edu2
50/lmhanson/page1.html**

Lichenland:
**http://mgd.nacse.org/
hyperSQL/lichenland/html/
biology/meeting.html**

The Living Soil—Arthropods:
**http://soils.usda/gov/sqi/
SoilBiology/arthropods.htm**

Microbe World:
http://www.microbeworld.org

Microbe Zoo:
**http://commtechlab.msu.edu/
sites/dlc-me/zoo**

Microbes—Invisible Invaders, Amazing Allies:
**http://www.miamisci.org/
microbes/facts18.html**

Mosquito Information:
**http://www.mosquito.org/
mosquito.html**

Soil Organisms:
**http://www.smartgardening.
com/Soil_Organisms.htm**

Stalking the Mysterious Microbe:
http://www.microbe.org

Yucky Worm World:
**http://yucky.kids.discovery.
com**

index

First published by Franklin Watts
96 Leonard Street, London EC2A 4XD

Franklin Watts Australia
45–51 Huntley Street, Alexandria
NSW 2015

This edition published under license from
Franklin Watts. All rights reserved.
Copyright © Franklin Watts

Published in the United States by
Smart Apple Media
1980 Lookout Drive, North Mankato,
Minnesota 56003

Library of Congress Cataloging-in-Publication
Data

Ward, Brian R.
Microscopic life in the garden / Brian Ward.
p. c.m. — (Micro World)
Includes index.
Summary: Describes the various kinds of
microbes found in a healthy garden, whether
helpful or harmful.
ISBN 1-58340-473-2
1. Garden ecology—Juvenile literature. 2.
Microorganisms—Juvenile literature. [1. Garden
ecology. 2. Ecology. 3. Microorganisms.] I.
Title.

QH541.5.G37W27 2004
577.5'54—dc22   2003058967

Editor: Kate Banham
Designer: Joelle Wheelwright
Art direction: Peter Scoulding
Illustrations: Peter Bull
Picture research: Diana Morris
Educational consultant: Dot Jackson

The publishers would like to thank the following for
permission to reproduce photographs in this book:
Frank Blackburn/Ecoscene: 9t. Kevin Brown/Holt
Studios: 18t. Dr. Jeremy Burgess/SPL 23, 27t, 29t,
Nigel Cattlin/Holt Studios: front cover tl, tr & br, back
cover tl & cr, 2b, 3t, 9, 12t, 13t, 14, 15, 16, 17t
20b, 22b. CNRI/SPL: 2t, 8. Dr. A.R.
Crooker/Custom Medical Stock/SPL: front cover bl,
3b, 11. Alan & Linda Detrick/Holt Studios: 18b.
Nick Hawkes/Ecoscene: 3c, 24t. Holt Studios: 17b.
Brian Knox/Papilio: 21.
Rosemary Mayer/Holt Studios: 4c, 4b, 7t, 22t. Phil
Mitchell/Holt Studios: 10c. Robert Pickett/Papilio:
25. Kjell Sandved/Ecoscene: 19. Saturn Stills/
SPL: 28. David M. Schesler/SPL 29b. Lee D.
Simon/SPL: 20t.
Laura Sivell/Papilio: 12b. Sinclair Stammers/SPL:
26. John Walsh/SPL: 24b. Robin
Williams/Ecoscene: 10b.